YOUR LIFE WILL BE DIFFERENT 31 DAYS FROM TODAY

Like anything we do at Maize Quest, this book is an experience—a 31-Day experience. This book is a crucial part. Just because the world seems to be going digital and paperless, we still value analog tools, regular pen and paper.

You can connect to the online component of the program at www.31DayTurnaround.com. When you connect with us there, enter your email and mobile phone number and the date you are starting your 31-Day Turnaround. We will send you inspirational daily quotes that correspond and support each day's Turnaround Focus during your 31 days.

You may also download printouts specific to a few of the days, a calendar to track your progress (just like Jerry Seinfeld uses), and exclusive BONUS materials.

Go to www.31DayTurnaround.com BEFORE you start.

THE 31-DAY WORKFORCE TURNAROUND

ONE MONTH TO A **HAPPIER** YOU AND A MORE **PRODUCTIVE** CREW

HUGH MCPHERSON

Copyright © 2018 Hugh McPherson

All Rights Reserved

Year of the Book
135 Glen Avenue
Glen Rock, PA 17327

No part of this book may be reproduced or transmitted in any form or by any means, electronic or mechanical, including photocopying, recording or by any information storage and retrieval system, without written permission from the author.

ISBN 13: 978-1-945670-78-7

ISBN 10: 1-945670-78-9

Dedication

This book and the personal transformation I continue to pursue would not have happened without the example of forgiveness from my father Paul, the irrational belief in me from my mother Gail, the decades of encouragement since the beginning of Maize Quest from Michelle, counsel from Matt—my sounding board for ideas big and small, and the love and encouragement of my beautiful wife, Janine, who once, long ago, in exasperation while proofing a speech for me said, *"If you want it to sound better, write it better!"*

INTRODUCTION

Do the math. The average 39.2-hour-per-week worker, in 50 years of employment, will spend over 92,000 hours at work. If you are a business owner or a farmer working 60 hours a week, that number climbs to over 150,000 hours at work. For most of those hours you won't be alone. You will spend them working with employees. For much of it, leading a team.

For so many business owners, working with employees feels like a constant struggle between 'us' and 'them'. So powerful is the memory of *'that one time when an employee stole money from me'* or *'when she called off on our biggest day of the year and I had to make donuts myself for 12 hours'*, that employees are forever adversaries—not to be trusted. Us against them.

What a terrible way to spend the next 100,000 hours of your life. What if there was a different way to interact with your people? What if you could build a team that was enjoyable to work with and got plenty of work done? What if your team wasn't a mindless league of robots, but started thinking of more efficient ways to serve clients, complete tasks, and innovate?

There is a way, and the process starts with you. *Today.*

This 31-Day Turnaround program is not filled with decrees from high atop an ivory tower at an ivy-league institution. If you want to read research papers from tenured professors, you must look elsewhere. This book is the latest chapter in a life spent at all levels of family business.

I started work at age 5 in our peach packing house at Maple Lawn Farms in New Park, Pennsylvania. It was 1980 and my parents, Paul & Gail McPherson, were working flat-out. My childcare up to this point had been my parents placing me in a peach bin—that's a farmer's version of a play pen—where the girls sorting peaches could at least keep an eye on me.

At age 5 however, I could get out of the bin, so they set me to work putting boxes on the box filler. Every 35 seconds, a box of peaches was pneumatically ejected and I'd put in another empty for 4-8 hours per day. It was my first exposure to working on the crew.

Over the next 13 years, I progressed from handing tools to the old guys to working on the teenage peach thinning crew, and eventually to leading the crew with instructions from Dad. All the challenges of being *the boss' son* applied. Summers through college I'd return home, and max out the hours I could work, so I wouldn't need a job during the semester.

Upon graduation, I received a good paying job selling insurance, and began my life as an employee. Those steady checks were great, but I entered a world of bureaucracy so entrenched that in my 9 months with the company, I couldn't even get a brochure printed to hand to prospects at trade shows. By the end of my time there, I would log over 2,000 cold calls, essentially the entire database, yet fail to sell a single policy. It was disheartening. I felt unsupported despite my effort.

All the while, every evening and weekend during 1997, I was working to create the first Maize Quest Cornfield Maze Adventure. Though it depends who you are listening to, we created one of the first ten corn mazes, ever. The full story would fill another book, but when I asked my parents what they thought of the idea, their answer was, "It's a great idea. Go write your business plan, incorporate your own company and get your own loan."

1997 was the first year running my own business. I was the boss! Then from 1997 to 2009, I ran the business, hired, fired, and managed employees the same way I assumed it should be done from my past experiences, from a few books and from seeing 'bosses' on TV.

Basically, I had no plan, no structure, and nothing but headaches. We might have been growing, but it wasn't working. I didn't think I had time to work on 'developing my team' until one day when we made a little girl cry.

My manager at the time and I were doing employee interviews for the summer season

of 2009. By 'doing interviews', I mean we were following a script I'd prepared during college interview prep. We were aggressively questioning, *"What do you bring to the table? Sure, your brother worked here, but are you as hard a worker as your brother? What would you do if a customer was yelling at you in the market?" and on and on and on...*

To her credit the 14-year-old girl held it together until she made it to the stairs leading away from the office, but then we heard her break down and start to cry. What had we done? *While interviewing for the Maize Quest Fun Park, we'd made a little girl cry?!*

Something inside me snapped. It was a moment of realization that everything about the interview process I'd been taught, all the preparation I'd done for interviewing, all the smart people who had interviewed me... it was all wrong. Overnight, I completely re-wrote the way we would do interviews, prepared a pre-interview sheet we still use today, and changed the way we perform this process forever.

That change in our interview process launched my desire to adapt the entire system by which we manage employees. It led within a year to my collecting and writing down all the systems, checklists, 'What Should I Be Doing' lists, job descriptions, and our home farm's linchpin–the Employee Expectations Guide.

When other farmers found out about the system we had in place and the difference it made, we codified it all into our Agritourism Manager Boot Camp program that has been delivered to over 120 farms as an online course.

The key was that I was trapped in my old mindset until that interviewee's tears broke down the facade. Her tears created the pain inside me that drove me to find a different way, to *make* a different way.

Something is hurting you, too. You picked up this book because you want things to be different between you and your staff. The only question is... *what hurts?*

This will not be a passive reading experience, and your first task is to write down, right now in the blanks below, what hurts most right now in your work life. In your relationship with your staff? What do you most wish you could change?

My staff coming in late drives me crazy!

How to use this book to begin *your* 31-Day Workforce Turnaround

Start on a Monday. This book is a daily, practical guide that will take you as little as 5 minutes per day. We recommend **you read each day's Turnaround Focus *out loud*.**

Sure, that sounds weird, but there is a different connection made in your brain when translating written words into speech. Essentially, when you read, speak, and hear the words, it grants the message *three times the sticking power.*

Read the brief article or story that accompanies the Turnaround Focus, and complete the Challenge for the Day by writing your answers in the blanks provided.

If you only work during the week, you may skip the weekends and follow the program over 31 'Business Days'.

In 5-10 minutes each day, you are empowered to make small changes in your behavior and in your interactions with your employees. The cumulative effect over 31 days will be *remarkable.*

Stick with it. Surely you can do anything for 31 days. If you miss a day, just jump right back in and continue from where you left off.

If this book sparks positive change in your business life and you want to continue with a complete system for employee management, check the back of the book for more information on our full Manager Boot Camp online program.

There is a different way to live the next 100,000 hours of your work life. There is a different way to work with your team more collaboratively.

It all begins with you. *Today*.

Day 1:

The Road to Success is Paved with Good Intentions

I fully intend to change the way I interact with our staff. I will approach every staff person with the intent to help, encourage, and serve them, so we can make the business better together. I won't be perfect, but I plan to get better and better one day at a time.

We remember shocking events vividly. The day your child was born. The day you were stranded on the road in the middle of a snowstorm. The day you landed a big client. The day you discovered someone was stealing from your store. The roller coaster ride at the amusement park as you started down the first drop.

Yes, 'shocking' is easy to remember. Life, however, doesn't happen in big shocking events one after another. Life happens one day at a time, one hour at a time, one minute at a time. Life is more about routines. That's why we remember, often even seek out a big 'shock'.

Routines are our mind and body's way of 'putting decisions on autopilot'. This is a great plan for the majority of our daily life. Do you really need your full attention for the drive you've made 100 times to work?

So what prevents a 'life shock' from changing every decision you ever make from this day forward? What prevents your life's routines from numbing you into complacency?

Your intentions.

The saying, *"The road to hell is paved with good intentions"* should be completed *before* repeated: *"The road to hell is paved with good intentions that were never acted upon."*

Your intentions are incredibly important, and your intentions with regard to working with your staff are crucially important as well. We never *intend* to insult our co-workers. We never *intend* to crush someone's new ideas. We never *intend* to berate a staffer in front of his peers, do we? It just happens...

It happens because we didn't have any intentions *at all* regarding how we were going to interact, until now.

Today's challenge:

Write your intentions below. If you agree with the intentions from today's Turnaround Focus, copy them here, then as you arrive at work, think through your intentions to change, to help, to encourage, and to serve your employees, your managers, and your co-workers.

I intend to be more patient with my people during training.

DAY 2:

DEVELOP YOUR LISTENING SUPER-POWER

I understand that the key to success is listening better, harder, and more intently than anyone else in the room. My people will not understand my thoughts and plans until they understand that I am listening to their thoughts and plans. I will listen without 'waiting to talk'. I will use listening to show respect and appreciation. I will cultivate the focus and skill of listening as a super-power to deepen relationships with my employees, my customers, and my family.

"It's listening, not imitation, that is the highest form of flattery." ~Dr. Joyce Brothers

You crave a listener. We've all struggled with the feeling that 'no one is listening' to us. We also know that the quote above is true: We desperately want other people to listen to us—when we rant, when we cry, when we win, when we hurt. Whenever we have an experience, our first instinct is to share, to have someone listen.

My wife and her college girlfriends are listeners. My lovely wife has friends from college we still get together with every New Year's, in fact they all refer to each other's families as The New Year's Gang. When these ladies get together for New Year's and a few other times per year, they download to the group. They talk and listen to each other often into the wee hours of the night. Talk and listen. Talk and listen.

My wife, typically quite early to bed, ends the late-night session energized, not exhausted. She has just had people listen to her. She has just been able to provide an attentive ear to each of the other girls. Those ladies have a connection that has lasted decades, through all the trials of family life and kids, because they listen to each other.

Don't you want connections like that? Don't you want a team that's in it for the long haul? Don't you want reciprocal relationships that energize you every day at work?

The key is listening. You must listen first. You must lead by drawing out ideas from your team and demonstrating respectful listening. You must enforce respectful listening during meetings, so the loudest voices allow the softer voices to be heard. You must champion listening by speaking directly about how you have failed at listening in the past, but are working to improve your listening skills.

Part of listening's super-power is that, when your staff knows someone is listening, they are more apt to contribute.

Today's challenge:

Begin to improve your listening immediately, today. When you are pulled aside by an employee or someone stops in your office, look away from your computer, grab a pad and pencil, look them full in the face and say, *"I'm listening."* Ask questions and dive deeper into the topic, even if you think you already know the answer or believe you know what the person is going to say. It may even feel painful to not instantly respond, but your goal is to build respectful relationships based on each person knowing that you will listen... *really listen.*

Day 3:

Light a Fire for Friday, Now

I will work to expand my planning to a week at a time. I know that making daily lists is valuable, but it limits my flexibility during the week. I will plan first the big objectives, then bring my team leaders in to plan the details for the week. I will plan appropriate amounts of time for tasks, engaging the wisdom of the team to make each work-week successful. I cannot accomplish everything alone. I cannot plan everything alone, but I will reap the benefits of team planning by week, not by day.

For years, I'd walk into work with very little planned and very little ready for the day. I've always been good at winging it, so we'd wing it every day. The trouble was, without planning ahead, we'd inevitably run into delays, missing supplies, too many workers, not enough workers. It was a train wreck.

The next set of years, I got good at planning the *day*. The night before we'd talk through what was going to happen the following day, I'd go run for the parts we needed at night, and we'd be able to make it through the next day. It was a big improvement.

My coach Jon Toy introduced me to starting the week with a clear plan for Friday, a vision for what would be done and how good it would feel when it got done by Friday evening. That's right—starting *Monday* with a vision for *Friday's* successful accomplishments. It's a game changer.

Why plan for the whole week? We all know that the 'best laid plans go awry'. If you are only ready for a single day, it can be hard to recover from a setback. Think about it. You have Tuesday slated for pruning, then it rains. You spend the day scrambling for activities, often busy work, for the crew.

If you have a week planned with both indoor and outdoor goals, you can move right from rainy Tuesday to start the indoor work slated for Wednesday. Your supply trips cover multiple projects for the week, instead of running each and every day. If you are killing it by Thursday afternoon, you can look ahead at Friday's work or just save the payroll by 'giving Friday afternoon off'.

Today's challenge:

Look ahead one full week. Picture how you would like to feel and what you would accomplish in a full week. Zoom your view out from the immediate day's tasks and expand it to one week. Plan your week and the big picture accomplishments, then schedule a meeting with your staff—just 30 minutes—to engage them in planning how, together, you can achieve your goals by the end of the week.

On Friday, we will have 4 blocks of apples pruned.

Day 4:

You Are Not Alone in Your 31-Day Turnaround

I recruit others to hold me accountable for completing my 31-Day Turnaround. I understand that having an outside accountability partner will spur me to complete this program. I understand that connecting to a group of professionals working towards the same goal will increase my chances for success. I so strongly desire to improve my relationship with my crew that I am willing to be held accountable for making real and lasting change in this program.

It's very nice that you have started this 31-Day Turnaround program. It's very nice that you say you want to improve your relationship with your people. However, are you fully committed to the process which is likely to be filled with struggles, distractions, and painful personal growth, or is this just another nice idea?

You need help to complete this program, but you don't have to believe me. The American Society of Training and Development (ASTD) study on accountability found:

The probability of completing a goal *(in this case, your 31-Day Turnaround)*:

- Hear an idea: 10%
- Consciously decide to adopt it: 25%
- Decide when you will do it: 40%
- Plan how you will do it: 50%
- Commit to someone else that you will do it: 65%
- Have a specific accountability appointment with the person you made the commitment to: 95%

– Source: American Society of Training and Development

Today's challenge:

Sign up online at www.31DayTurnaround.com and request to join our FREE Facebook group of professionals who are working through this program. ***Locate someone outside your organization, gift them a copy of this book, and have them start the program with you*.** Together, you are stronger, and when you set a definite date to connect with your accountability partner, you increase the rate of goal completion to 95%.

COMPLETE:

My accountability partner is ________________

and I'll reach out to them via ________________

by ____________________ date.

A great resource for finding a local mentor, coach, or accountability partner:

Society of Retired Executives

SCORE is the nation's largest network of volunteer, expert business mentors, with more than 10,000 volunteers in 300 chapters.

https://www.score.org

Day 5:

Tension to Manage or Problem to Solve?

I will identify and consciously differentiate between tensions to manage and problems to solve. I will permanently solve problems, and I will set in place systems to manage tensions.

Andy Stanley, the pastor at Buckhead Church in Atlanta and one of my favorite podcast ministers, asks, *"Is it a tension to manage or a problem to solve?"*

This is a key question you will use throughout your business life. As you work through business challenges, work to categorize tasks, processes, and management functions as 'tensions' or 'problems'.

Some examples.

If you are constantly running out of napkins in the snack bar, you have a problem to solve. Assign napkin ordering and estimation to a team member, coach them with your napkin ordering experience a few times, then you should never have to 'manage' napkins again—problem solved.

I was being dispatched by my team to run for parts, pick-up extra party supplies, and order pies. Maybe you have found yourself 'dispatched' as an errand boy or girl for your team. Why?! YOU are the MVP—the Most Valuable Player—and you are spending time driving to Wal-Mart?!

This was a problem. Solution: I got my two top managers their own Discover Cards, connected them to my account for monitoring, required receipts to be submitted, and told them to pick-up/order the supplies they needed. I haven't had a single problem, and I am no longer 'dispatched' as their errand boy. Problem solved.

In the agritourism business, we are always at the mercy of weather. With a 50% chance of rain, we must plan for a sunny day and lots of guests, AND we have to have a plan for showers and only 25% of the usual attendance. This is a tension to manage. You may have someone assigned to schedule staff. If you scream and yell every time you are short-handed, guess what? That person will schedule extra employees every day. If you stomp around any time you have a few extra employees during a lull in the day's flow, your scheduler will keep you perpetually understaffed.

Scheduling and managing staff on hand is a *tension to manage*. The system must be flexible enough to allow for a few employees to do things such as empty trash and sweep floors, so that you have them when the 1 PM rush arrives. Your process must understand the natural flow of the day, so you don't open with 20 employees, *which would leave too many on shift twiddling their thumbs before guests arrive,* but instead have 20 staffers by 4 PM during pumpkin checkout rush.

YOU must be flexible enough to understand that your scheduling person can't *'solve the problem'* of right-sizing the staff every hour, but must manage the employees on hand following a few guidelines you've put in place.

Some rapid-fire problems:

- All the doors at the packing house are left open each night.
- Chronically late employees.
- Can't find the tools you need when you need them.
- The Internet is too slow.

- Your POS System isn't giving you the reports you need.

Some rapid-fire tensions to manage:

- Growing conditions.
- Corn prices.
- Cash flow.
- Payroll levels during on and off seasons.
- Marketing budgets vs. weather events.
- Interacting with your teenage children.

Today's challenge:

What tensions do you manage? What problems can just be solved? Write down your daily or repeated frustrations. Categorize them, then adjust your approach and plan to handle them as a problem to solve or as a tension to manage. Discuss the difference with your staff so they understand your new language around problems and tensions.

Day 6:

Share the Big Picture

I choose to open up my vision, my big picture thinking, to my staff. I understand that opening up is challenging. Clear understanding of the vision for my business will activate in my people untapped energy. Their energy combined with mine will propel us further, faster, together.

This is a physical challenge. Stand up right now. Lift up one leg, hold your arms out to the sides, look up at the ceiling, then... close your eyes.

It's hard to keep your balance in the dark, yet so often we ask our people to do just that. We ask them to perform their tasks with no connection to the bigger picture.

We say, *"Sell, sell, sell!* *We've got to sell more this week to make things work out!"* Well, how much more? Does 'make things work out' mean I'm out of a job? Is the company going under? Are you going to miss payroll?

Or we say, *"We aren't getting enough done*. *You need to get that kitchen staff motivated."* Is the kitchen crew behind? How far behind? Are you really talking about the kitchen crew or the farm market in general? Are you frustrated and making a statement or are you giving me instructions? What do you mean by motivated? Should they be smiling as they work, or increase donut output?

Or we say, "Well, last week we just sold 100 pies." Does that mean we failed? Was that the store record? Do we normally sell 100 pies in a week? A 1,000?

How can your people 'get motivated' when *they don't know the goals?* Asking them to continue working without goals is akin to asking them to balance in the dark. We are essentially asking people to give 110% effort each day with no defined target. In its simplest form, it's quite gratifying to check off tasks from your list.

Do you work without targets? Without goals? Certainly not! You want revenue, sales, cost-savings numbers to hit or exceed your expectations, and think about how great it feels when they do. It's a gratifying reward just to 'hit your numbers'. It's a chance to celebrate.

Now, you do not have to 'open your books' to everyone in the company. Some businesses choose this path of radical transparency, but it is not necessary to reap the benefits of sharing the big picture. Let your people in on the greater plan and how as individuals they each are contributing to it, to the big picture, and they will bring new energy to build upon your own.

Today's challenge:

Write down each team member, where they fit into the big picture and to which goal or target of yours they connect. Once complete, schedule a brief time with each person or team to clue them in on their role in the big picture.

Jerry: Handles farm market supply ordering.

Day 7:

Close the Loop

I will focus on completing tasks and projects today. Before I begin any new projects, I will close one open project. I know that value is created by finishing, not by starting. I will allow my people time to close projects *before* redirecting them. I will build a culture that 'closes the loop'.

The joy, the fun, the energy is all in the beginning. When I have a new idea, I'm on fire for the new project, ready to take on the world as I kick-off my newest plan, product, or idea. I like to launch.

We've all experienced what happens next. We hit the wall. All that energy from the initiative's launch burns away and we're left with real work to do. Somewhere in the midst of the real work, we have a new idea! This new idea is surrounded by energy! Launching this new idea seems much more fun! So, we put aside our first idea and its drudgery to chase the new one. Now we have two ideas, two projects, neither completed.

Task switching is time lost. During a regular day, we often spend valuable time switching our crews from task to task to task. Maybe you are out with the crew pruning peach trees, when your driver comes back with a supply truck to unload. You load everyone up, drive back to the shop, unload the truck, load everyone up and get back out pruning. No more than 45 minutes go by, and your phone alerts you to a weather event coming in.

You really wanted to get the mulch down on the blueberries before the next rain, so you load everyone up and head to the blueberries. The crew waits there until you help your driver hook up the spreader, then you go back for the loader, get everyone started spreading. But after 25 minutes, the belt breaks, the rain starts, and you end the day with no pruning done, no blueberries mulched, a broken spreader, and a full payroll!

Can you see the inefficiency of switching? Had you left the truck to be unloaded later, left the blueberries for another day, skipped all that loading and unloading of the crew, and kept them focused on pruning, at least the pruning would be done!

Staying focused on *completing work in progress* and NOT SWITCHING actually gets more done and makes your employees happier; they get to stay 'in the groove'.

Today's challenge:

List your open projects, all of them. Repairing the deck on the pirate ship, researching a new spray controller, finishing that continuing education credit for food licensing, painting the market—everything. Pick the one that will take the least amount of time to complete, and close the loop. Then, and only then, move on to the next item on the list.

-Repair tour wagons

Day 8:

Today You Fall on Your Sword

I will lead with humility. I understand that success for my people and for my customers is success for me and that each employee's workday struggles are my responsibility. I don't have to be right all the time, in fact, I'm strong enough to be wrong, to be corrected and to change my stance if it is for the good of the company. I know that the good of the company means good things for us all.

It always takes two to fight. I used to have quite a temper. Being young, brash, Scottish, and fancying myself brilliant led to many inglorious, ego-fueled bouts with customers, employees, family members, and clients. Pride does indeed precede a fall.

This pattern of behavior didn't abate until Ed Staub, a Sandler Sales trainer, talked about *'falling on your sword'*. The idea is that before your employee can kill you, you fall on your own sword, leaving no adversary with whom the employee can fight.

Imagine you are speaking to a disgruntled employee. This staffer is quite upset and is increasing in volume. You do not agree with her assessment, so you begin to argue. Noting that you 'aren't getting it', she increases her tirade, tossing in some profanity. You mention how stupid this is, and consequently how stupid she must be, and on and on and on...

Or... Imagine you are speaking to a disgruntled employee. This staffer is quite upset and is increasing in volume. You do not agree with her assessment, but you understand that her perception is that your leadership has let her down. Instead you say, *"Oh, my. I am so sorry that I have let you down. Could you please explain the situation, so I understand exactly what's frustrating you?"*

After a brief stunned silence, the staffer begins, *"Well you sure did, so, uh, you see, it was so busy in here this weekend that the donut crew took all the flour. Now here I am with an order today and those people have left me short AGAIN!"*

You say, *"I'm so sorry to hear that. Yep, it's pretty busy. How can I help make sure you don't come up short for your order today? Let me help fix that right now, then let's meet with the donut crew Thursday when we're ordering to make sure we all understand how to get the order right for next week."*

Had you or your donut team done anything wrong? Nope. *Did it still cost you some strength to take the hit for the donut crew?* Yep. *Was she overreacting?* Yep. *In the big picture, did you make the right move?* Yep.

When treated with humility and kindness, people generally respond in kind. Your baker was upset because her responsibility for the new order was on the line. Essentially, she was upset because she didn't want to upset a customer. That's great! Exactly what you want in an employee! Likely, your donut crew was simply trying to meet the demand over the weekend—also great!

You, as the owner, must be accountable for both teams, keep them focused on the big picture, help immediately, then set up the system so neither crew is short materials again.

***Average* business owners are transactional.** They focus on the immediate transaction, which in this case was one team getting angry at another team. You are moving and growing to be an *exceptional* business owner, a *relational* business owner, one who sees beyond the immediate transaction, and in this case, sees a frustrated employee trying to meet her customer's order.

You see the big picture. You are willing to humbly fall on your sword instead of dueling with employees, or letting them duel with each other—an act that always ends bloody.

Today's challenge:

Be humble in your leadership today and find opportunities to lift up your people. Find an opportunity to fall on your sword so no one else gets cut. Hold yourself extremely accountable for every action as you operate to serve others, protect others, and grow your business by growing your people.

____

____

____

____

____

____

____

____

____

Day 9:

Trust Until It Hurts

I am daily increasing the trust I place in my employees. I know that to grow my business, I will have to give away control or face my own limitations of time, ability, and attention. I will place new responsibilities carefully and with adequate time for training each person to succeed. I will increase my team's decision-making authority to include small expenditures and details that work towards business-wide goals. I will make sure my people have a clear picture of each project to which they contribute, so they are empowered to be successful decision makers.

You are the limit. Many business owners *want* to be the limit. They only want the business to grow to the extent at which they can be in control of everything. They make all the decisions, talk to all the clients every single time, and they alone do the actual work or make the actual product. That's fine, but it is very limiting.

As soon as you add a single employee, you must begin to let go, to trust. We can all think of one-man electrical contractors or one-woman marketing firms. That business owner is in complete control. In our businesses, we can't even open the doors, let alone staff a Pumpkin Patch, without help. We *require* help, which means we are *required* to trust other people.

Trust and mistrust. If you've been on this Earth beyond infancy, you can remember a time when you trusted the wrong person, a time when your trust was betrayed. From a group project in school when one kid didn't do her part, to employee theft from your registers, chances are you have good reason to be mistrustful of other people, *right?*

Wrong. Because the fear of loss is so powerful, we tend to exaggerate the negative, the instances of mistrust. We often discount the positive interactions we have with trustworthy people on a daily basis. It is this fear-filled mindset that inhibits our ability to trust people and it limits our potential for growth.

C. McNair Wilson, former Disney Imagineer, encourages us to "Assume Brilliance®". This means that we go through life looking for all the ways people are behaving well, being helpful, acting in a trustworthy manner—we are to look for the positive, the brilliant in everyone. We are to trust them.

If you have people you can't trust, you need to let them go. If you are unwilling to let people go, then you must be willing to trust them. So, do it.

Today's challenge:

Find one task or process today that you can start handing over and entrusting to one or more of your people. Train them, do it with them, then entrust them to do it. Repeat and repeat until it feels as though you are giving it all away... until it hurts.

Day 10:

Get Your Brain Off the Treadmill

I will stop trying to remember everything and write it all down instead. I no longer entrust my success to my ability to remember. I will capture my brilliance by emptying my mind onto paper.

Have you ever been driving and had a flash of brilliance? An idea that could change everything, then you are at your exit, check the traffic light, proceed through the intersection and suddenly realize you can't remember your idea? It's maddening.

Today is a practical application day. David Allen, in his book *Getting Things Done*, states that our brain's short-term memory can hold no more than 7 to 12 items.

When we leave something undone or suddenly remember something to do, or think of a new idea, our brain scrambles desperately to remember all that we put into it while we do our work, handle customers, or drive a car.

The brain is exhausted by this struggle to keep everything in our mind, but, says Allen, by writing down each task, step, idea, and grocery item on paper, our brain can release and relax. You can get your brain off the "Treadmill of Remembering".

Buddhists call this constant distraction of trying to remember the 'Monkey Mind'. They refer to it as the noisy thoughts that rush in when one tries to meditate. It is the avalanche of thoughts your mind is trying to remember.

The only way to overcome this Monkey Mind is to empty it out onto paper. Writing things down is the only relief your brain understands. Once on paper, your brain can relax, sure that it won't be forgotten.

Mind mapping is a brainstorming process by which you write the main topic, for instance 'Farm Market Improvements', in the circle in the center of a sheet of paper, then absolutely anything that comes to mind, you write down in circles that connect, like a map, extending out from the center topic.

Each topic branches off into sub-topics. The goal is to write as much as you can, as fast as you can, with NO EDITING. The goal is to dump everything out of your head on this topic.

Once you have completely emptied your mind on that topic, do a new map for each major area of your business or project. Take a break from the mind maps for the rest of the day.

The next day, revisit the maps to see if you have anything left in your brain to map out. If not, begin to look through the maps and create lists of tasks for each sub-topic of actionable items, then delegate what you can and get to work.

Often when I feel overwhelmed or get that 'I don't even know where to start' feeling, a round of mind mapping will empty and clear my mind, allowing me to see clearly the next step toward my goal.

With your brain off the 'Treadmill of Remembering', you'll likely feel a new freedom and positive sense of purpose.

Today's challenge:

Create a mind map for one of your most pressing challenges or one of your most interesting ideas. See where the map leads you and enjoy the relief of an emptied mind.

Space for Your Mindmap:

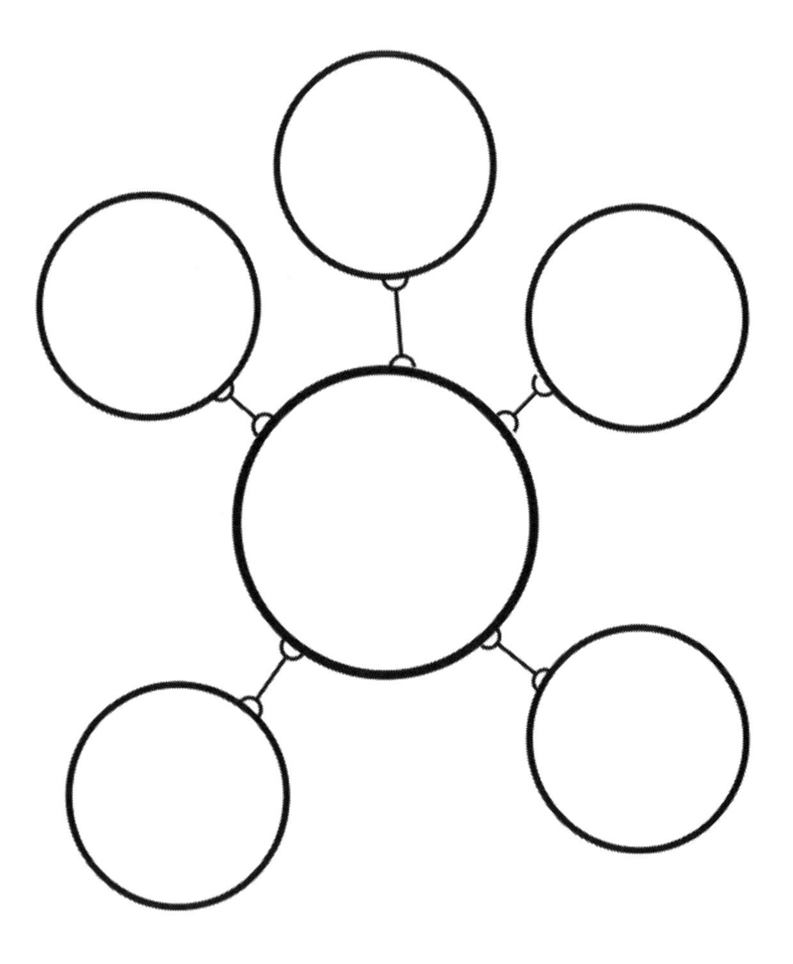

Get more templates and sample mind maps at www.31DayTurnaround.com

Day 11:

Accountability Partners

I understand that I need help to create and maintain lasting change. I will purposefully engage an accountability partner in my goal to improve my relationship with my staff. I will connect with my accountability partner, and explain this book and my plan for self-improvement to that partner to keep me focused throughout my 31 days.

Self-motivation will only take you so far. As the leader, you need someone from outside your business to hold you accountable to your goals and plans. I felt pretty silly when I thought about getting a business coach, so I hedged my bet. I joined a Mastermind group. I had been re-reading Napoleon Hill's classic *Think & Grow Rich*, in which he talks about accountability and the 'Mastermind Alliance'. Basically, you need someone else holding you accountable.

Small business owners are the owners, the leaders, the boss. We do not generally have a superior in the business who can be our mentor or hold us accountable. Often, it is this freedom that attracts us to business ownership—we want to be the boss. Unfortunately, it comes at a price: We must internally develop all the motivation we need to succeed each and every day. We are alone.

I was reminded that LeBron James has a coach. Lionel Messi has a coach. $100 million baseball players report to a coach. What on earth makes business owners like you and me think we DON'T need a coach, that we don't need help to be our best?

So I hired a coach. I reported to Jon Toy that I wanted to write this book and he provided not only encouragement, but contacts for the editor and publishing tips from writing his own book that I needed in order to bring it to life. Getting accountability help, getting a coach, getting a mentor, recruiting someone from outside your family and organization might be the very thing that spurs you to success during your 31-Day Turnaround and beyond.

If you don't have coaching resources around you, plug in to our online Facebook VIP Group for this book and access a wealth of experience from other business owners just like you. Go to www.31DayTurnaround.com and opt-in for an exclusive invite. This group is NOT open to the public.

Today's challenge:

Recruit an accountability partner to help you through the 31-Day Turnaround program and beyond. It doesn't have to be a professional business coach, it can be a colleague or fellow business owner, but it must be someone you respect, trust, and with whom you can share your goals and aspirations. Start looking today.

____

____

____

____

____

____

Day 12:

Delegate Responsibility and Authority

I will delegate and entrust responsibilities to my staff along with the full authority to get the job done. I willingly give away some of my personal power so that I can empower my people to confidently act on my behalf. After I delegate, I will coach, but I will not allow tasks to return to me. I will spend the time training my people so they are comfortable with their new responsibilities and authority.

Have you ever... handed off a task to a staff member, then when they struggle, they immediately come to you to 'fix' the problem? You finally delegated a task and then left the job site to do your own work, only to be called on the phone or radio a dozen times with a dozen little questions or decisions to make? You delegated only to find yourself actually doing the job for your employee?! AND they are standing next to you watching you do it?! How did that happen?

Monkeys. That's how. Monkeys. To be clear, the 'monkeys' in this metaphor are the tasks, not the people.

Why does your staff keep trying to put monkeys back on your back? 3 Reasons.

- **You've trained them to be helpless without you.** The classic way to do this is to yell and criticize your people after they complete tasks. If you seldom praise your people for jobs well done, they will think the only way to avoid being yelled at is to involve you in every decision. That way, "It's just the way the boss wants it and I won't get yelled at."
- **You are too good at everything.** If you grab the job back from your people the second they have even the slightest difficulty, they know they can let you do all the work all the time! They think, "You know, if I just call Bill back here and say, 'This mixer is giving me a fit', he'll come back and make the next 4 batches of donuts..."
- **You won't teach them anything.** Are you the only one who can operate the grain dryer because

"it's too complicated for anyone else to possibly understand!" Well then, by golly, you'll be the only one ever operating the grain dryer, so you might as well chain yourself to it!

To avoid monkeys coming back:

Assign monkeys correctly. Picture the job you just assigned to your staff person as a monkey. You had it to do before, but you handed it off, and the monkey jumped from your back to your staff person's back. To ensure that the monkey stays where assigned, you must:

- **Give clear and thorough instructions, including training** if the job is NEW to the employee.
- **Give a clear deadline when it should be done.** (This is tricky, not the deadline in which YOU could have done it, but a realistic deadline for a staff person without your full experience and motivation.)
- **Give pointers, and answer questions with questions.** Inevitably, the employee will ask questions, or ask for decisions to be made by you, or imply that you should come and do it for them. Reply patiently with questions such as, "What do you think I'd say to do?" or "What would likely be my answer to a guest in this situation?" or "If I weren't here and you had to get it done, what would your plan be?"

Finally, once the job gets done:

PRAISE the good work done without you.

PRAISE decisions made without you.

COACH through items or decisions that you might prefer to be made another way, or make a checklist for the next time, but don't lay on a big critique.

The goal is to not have to do the task again or be bothered during the next time the task is underway. Keep the monkey—the task—in the right place, with the right employee, and don't let it back on your back.

Today's challenge:

Find a task, a monkey, that keeps returning to your back. Determine who among your staff should be responsible for handling the task, then sit down with the person and review the process for completing the task. Are they missing resources? Tools? Confidence? Training? Help from other workers? Funding? Give away the power to make decisions to your staffer and move into the role of coach. Once completed, find the next monkey and repeat.

Document the steps as you go – *don't miss tomorrow's challenge to learn more on this!*

Office supply ordering / to Jenny / Staples.com login info & list of weekly supplies to stock.

Day 13:

Procedures Beat Magic

I will examine our work to build a library of procedures that enable my team to function without my direct supervision. I know creating and writing procedures takes time now, but it will save time every day moving forward. I choose to invest my time now, to save countless future hours, and make my business run correctly every time, for every customer.

"It's our people that make the difference."

"Customer Service is our competitive advantage."

"Our farm market people are the best."

"Our CSA box packers are magical."

Magical People. It sounds so good, doesn't it? You hear it all the time: 'No one can beat our people'. You also often hear, 'I can't find good people'. Sometimes you might think to yourself, 'I could never replace Linda because no one seems to be able to do what she does'. In this case you are both enjoying and suffering from, 'Magical People'.

Magic does not equal healthy. It's fine to have Magical People helping you run your business, but it's not healthy. In fact, it is remarkably unstable. If a Magical Person gets sick, you're in trouble. If you want a vacation and you are the Magical Person, the business is in trouble. If you want to grow your business, you can... right up to the limit of what you and your Magical Staff can possibly handle in tasks, phone calls, tours, and exhaustion. Depending on 'Magic' means you will have severe liabilities.

When Magical People move on. In my life on the farm we've lost 5 indispensable people, but the farm's still here. My dad is 76 years old and is absolutely Magical, but we'll be replacing him within the next 20 years... almost for sure.

False belief in Magical People. Sorry to burst your bubble, but Magical People don't exist. The sooner you accept this, the sooner you can move on from your childhood belief in Magical People. It's time to pull back the curtain and see what's behind those magical tricks.

Behind the curtain. What you'll find is that the only 'magic' people bring is the ability to create good operating procedures and behaviors to get themselves through the day and the work done successfully. That's it.

Don't for a minute think that I'm raining on the parade of magical actions you've taken, or magical sacrifices you've made. You are magnificent! It's just time to move on and grow your business, and build your team from a group of wannabe magicians who wonder daily, "How does she do it?" into a team with full access to your bag of tricks.

Decoding the real magic is in the process. A 'Magic' example: Why can no one get the farm market displays to be as beautiful as Sherry?

Decode Sherry's Magic:

- Have Sherry set up the farm market jam display and make it beautiful.
- Take a picture of Sherry's incredible display.
- Write a verbal description of the display and how to restock it to look exactly like the picture.
- Laminate the picture and hang it on a nail under the display or put it in a binder of Market Displays.
- Move the stock for that display to a location directly underneath the display for easy refilling.

- Hold a training session for anyone who restocks the market to explain the procedure for each display.
- Set the expectation that the displays shall only be restocked to match the pictures.

Magic decoded. If Sherry from the example above is absent—no problem! If Sherry is the owner and 20 people work in the market—no problem! You now have a procedure instead of a magical person, and discipline or correction is easy—the display either matches the picture or it doesn't.

Build systems using the brilliance of your magical people, then run the systems with the rest of your team.

Today's challenge:

Find one task or job that you think is done by someone, maybe you, who you think is irreplaceable. Break the complex task down into step-by-step instructions. Add pictures as appropriate to ensure the end product is exactly the way your magical person would do it. Coach to the procedure to ensure you are holding every employee to the same standard.

Making our world-famous donuts: Measure 3lbs of flour, heat cider to 72degrees...

____

____

____

Day 14:

Acknowledge Your Failures, Then Burn Them

I know that I have failed my staff. I have lost my temper. I have belittled them publicly. I have dismissed their input. I have left them untrained to handle challenging situations. I acknowledge my failures as a manager. I also acknowledge my willingness to improve myself and my interactions with my staff. My failures in the past will not define my future as a manager.

You have failed, but you are not a failure. There is no tougher critic than our own internal one. No one can be tougher on us than we are on ourselves. When you look honestly at the past, you have failed many times as a manager, but failed attempts do not define us. You define who you are each and every day.

The fact that you are working through a book called *The 31-Day Turnaround* indicates your personal admission that you aren't as good as you want to be. It indicates that you are wise enough to seek tools to affect change in your life, and it indicates that you are strong enough to create change in your own life.

Acknowledge your failures, then burn them. We were visiting State College, PA—home to my alma mater Penn State University over New Year's. In the midst of the park were ice slides, sculptures, food vendors, and a curious pavilion with fire pits. On tables around the fire pits were small slips of paper and markers.

Each reveler was encouraged to write down a hurt, a struggle, a tragedy, a loss on the paper, then toss the hurt, the shame, the pain into the fire. It is a symbolic gesture, but one that captures the ephemeral nature of the past.

"I tell you the past is a bucket of ashes, so live not in your yesterdays, not just for tomorrow, but in the here and now."
~Carl Sandburg

Today's challenge:

Take slips of paper, and on each write a separate failure from the past. As you feed the slips (safely) into a fireplace or pit, say each failure out loud—then let it burn.

EX:

I really over-extended my budget to get my new truck and I've been paying interest ever since.

When Michelle lost the last order forms, I chewed her out in front of everyone.

Day 15:

Safety Builds Security

I make the safety of my people my top priority. I know that if my people are worried about their personal safety, they cannot effectively do their work. I will budget for safety and immediately correct unsafe conditions. I will demonstrate through my workplace actions a no-tolerance policy for unsafe practices, even if the job takes longer to do safely. I know my commitment to safety demonstrates my commitment to my people.

When I was a kid, we had a tractor with a PTO shaft, that's Power-Take-Off shaft for the non-farmers, that never really stopped spinning. This PTO shaft is the way power was 'taken off' the engine and directed to mowers or pumps. Now, we could have backed up to the mower to hook up, turned off the tractor, connected the PTO shaft, then restarted the tractor, but the tractor didn't always want to start. So we'd leave it running, grab the PTO shaft, and wait, wait, wait, SLAM it onto the tractor as it slowly spun. It's amazing I still have my fingers and toes.

Personal Safety is at the foundation of Abraham Maslow's famous hierarchy of needs, because it is the base from which all other needs are built. Quite simply if you don't feel safe, you can't get anything else done effectively because you spend too much time worried about your safety. Self-preservation is hard-wired into our brains.

It's wired into your people as well. If they are working in unsafe conditions, Job #1 is their personal safety, which means that whatever you have them doing is Job #2. By eliminating their concern for their personal safety, you will make the work Job #1 again and in turn make them more efficient.

This has another positive effect: strengthening bonds between worker and workplace. By demonstrating that you care enough for your employees to protect them from harm, you are showing that you care for each staffer as a person. People don't work for companies, they work for people. They prefer to work for people who care for them.

Today's challenge:

Evaluate your equipment, your mixers, your tractors, your drill press, your ovens, your forklifts, your railings, your ladders. What safety concerns can you easily eliminate? Are you short of chemical gloves, safety glasses, ear plugs, dust masks, or painting coveralls? Are the guards off your mowers, chainsaws or weedwackers? As you go through the day, note when you see your people not protecting themselves and gently correct the situation, noting your commitment to their safety.

Order locking hitch pins & ear plugs.

Patch hole in hay barn floor.

Day 16:

What ONE Tool Can Make Your Life Easier?

I will specifically ask each of my employees over the next few weeks, 'What one tool or person or task or use of my time will make the rest of your season easier?' While I think I know what my people need to do their jobs, and I think I know what I would use to do their job if it were mine, I will express my desire to listen and know each person better by asking for their 'ONE tool'.

You think you are so smart. We all do, and we are in our own world. As farmers, we have to know so much about so many things that by the end of 15-20 years in the business... we are really smart.

As our business grows, however, more staff comes on board—**staff members who are specialists**. To use our work time efficiently, we must delegate tasks to those new specialists and let them help us get more done. In the process of delegating, we give away some of our power and control, trusting that the specialist will get the job done for us.

I call it 'drift'. Drift is the ever-increasing distance in terms of actual work, and eventually, even the understanding of the work being done that naturally occurs when we delegate. I used to be mildly helpful when working on equipment, but I delegated that to my shop mechanic so long ago, the drift has completely separated me from the job. I no longer know how to work on the tractors because it is no longer my responsibility.

So today, I am completely separated from the work my mechanic does on a daily basis. This is great, because I have other things to do! It also means I have very little understanding of what he does, and specifically what he needs to complete his job. If I walk into the shop and tell him what order to do his tractor servicing tasks, he'll be pretty upset.

If I walk into the shop and ask, *"Hey, Frank, is there ONE tool that would make all this tractor servicing a whole lot easier?"* I've instead set up a positive situation, one in which I'm *listening* to Frank (remember your '*Listening Superpower*' from Day

2?) and asking for *his* input to make *his* life easier. Sure, you might get an answer such as, *"A new shop with marble floors,"* but often those answers are just the staffer seeing whether you are really listening. *With a few tries at it*, you might get an answer that a tire machine to pop tires off rims would be the best, because it's "really a struggle to pop tires with the old hand bars." Then, see if you can find a machine within your budget at a sale.

Listening is the key. Andy Stanley, minister at Buckhead Church, tells the story of asking about the 'one tool' at a youth ministry meeting and discovering that the 'one tool' was a paper cutter for the youth materials. That was it. *A paper cutter*. In our office, Jenny just needed her own P-Touch label-maker, then she saved hours of time running the other label-maker back and forth between buildings.

Today's challenge:

Find one employee today to ask, *"What's the ONE tool that would make your job a lot easier?"*

____

____

____

____

____

Day 17:

Separate Urgent and Important

I will focus my team on urgent, important tasks through a carefully devised weekly plan. My time will be focused on important, but not urgent tasks and planning so that my time is used to prevent emergencies. Fighting fires and being in a constant state of emergency is not healthy or effective. I will work towards spending all my time on important, but not urgent tasks.

Urgent/Important, Urgent/Not Important, Not Urgent/Important, or Not Urgent/Not important. Dwight Eisenhower has been ascribed the Eisenhower Matrix for time management (depicted on the next page). The quote that leads to the matrix was, *"I have two kinds of problems, the urgent and the important. The urgent are not important, and the important are never urgent."*

1) **Urgent/Important:**

   Client needs materials delivered before the weekend.

2) **Urgent/Not Important:**

   Your email notification just 'dinged'.

3) **Not Urgent/Important:**

   How will your farm attract the next 3,000 guests to grow?

4) **Not Urgent/Not important:**

   There is filing to do for your monthly receipts.

Today's challenge:

Use the matrix on the following page to classify your list of To-Do items. Throughout today, note your activities and classify them in the appropriate box in the matrix. In which box do you most often operate?

	Urgent	**Not Urgent**
Important		
Not Important		

Get more templates at www.31DayTurnaround.com

Day 18:

Close the Rings

There are three things so important that I will do them every day. My three tasks each move me closer to my goals. Like rings on the Apple Watch, I complete the tasks and close the rings every day as a positive habit. I celebrate closing these positive habit rings by the end of each day.

Your habits will make you or break you. *"We are what we repeatedly do. Excellence then, is not an act, but a habit,"* says Will Durant, author of *The Story of Civilization*. It is easy to get wrapped up in waiting for that 'one big thing' or your 'ship to come in'. It is the idea that everything in your life, all your troubles will be solved in one great stroke of luck.

The world of "if only." If only my finances were fixed. If only I could find the right man. If only my wife were more supportive. If only I'd win the lottery. If only it wouldn't rain in October.

You might have wished some of these very wishes yourself! Living in the world of 'If Only' places you in the victim mindset—you are watching the world happen to you. You are no longer in control.

You don't need a smart watch to make this happen. On the Apple Watch, the 'rings' on the screen represent "Movement', 'Exercise' and 'Stand'. To 'close' these rings and have a healthier day, you must log enough steps, increase your heart rate for a set period of time and stand-up long enough from your desk. Three rings, three behaviors, one goal to be healthier.

Take back control with 3 small tasks. Today you are going to begin a new habit by selecting 3 small tasks you can complete each and every day that will slowly work over time to improve 3 areas in which you struggle.

Not getting along with your spouse? Commit to giving one genuine compliment per day.

Trouble with cashflow? Commit to viewing your accounts once per day.

Trouble with exercising? Commit to one pushup per day.

Trouble with workaholism? Commit to 15 minutes talking with your spouse about anything but work once per day.

Need more clients? Commit to one client phone call and one Facebook Business post per day.

Trouble with your staff? Commit to saying “Thank You” once per day.

Teenage children getting out of hand? Commit to one family [or Daughter/Dad/Mother/Son] meal [or at least snack or hot chocolate] together per day.

The first few times, it will feel like nothing is happening; nothing is improving. Over time the people affected by your very small new habit will see a preponderance of evidence that you are dedicated to improving yourself and your relationship with them.

Today's challenge:

Pick your 3 biggest areas of struggle in your life, work, family, or personal development. Write them down. Choose one very small task to complete each day. Set reminders on your phone, put sticky notes on your computer. Tape them on your dashboard. Stick to completing the small tasks each and every day for a two-week trial period, then check your progress and switch them up or continue using them as your three rings.

Work

Talk to a wholesale client ______________________

____

____

____

Family

Offer one compliment per family member __________

____

____

____

Personal

Take a brisk walk through the farm ______________

____

____

____

Day 19:

Where You Spend Your Time

I will track my time and create an honest assessment of my habits to better use my most precious resource. I will account for my time in 15-minute blocks for a full day to understand how I spend my time. The better I understand how I use my time, the better I will understand how to allocate my people's time. When I understand my time-based challenges, I can coach my team to greater efficiency.

You won't believe where you spend your time. My coach Jon was the first to convince me that I needed to track my time down to 15-minute intervals. This is an eye-opening experiment. To help you complete this pain-filled exploration of your work life, we've included a FREE Time Tracking Template at www.31DayTurnaround.com.

Get a clipboard. Tie a pen to the clipboard, mount the Time Tracking Template, and set the clipboard at your bedside for the following day. When you wake up, enter your first log after your first 15 minutes awake. Did you walk the dog? Take a shower? Make breakfast? Read the newspaper? Pack lunches for the kids?

Then what? Set a repeating timer on your phone and honestly enter your activities 15 minutes at a time. *Did you commute? 15 minutes, 1 hr 15mins? When you get to work, how long did it take you from car to desk, to coffee, to chit chat, to boot your computer, to answer email. Ooops!—you were on Facebook for how long?!*

Track your home/family time. *How long did you help with homework? Cook dinner? Run kids to activities? Talk with your spouse/partner? Watch TV? How many minutes did you spend on your phone? (They have tracking apps for this one—I use RealizD.)*

The more you resist, the more you need it. You are going to hate this task, but know that the more you resist, the more you need it. Your resistance is directly proportional to the amount of time your brain knows is subconsciously wasted.

You need this check-up. You need this accountability.

This Time Tracking Template often has two results:

#1 You find out that you have been wasting a lot of time on frivolous activities. When you say, "I wish there were more hours in a day," you are likely going to find out that you have plenty of hours, you are just mis-using them.

#2 Simply because you are tracking your time, you are going to behave differently… more effectively. That which gets measured, gets improved. You know you are tracking time, so you are likely to put in some of the most effective days when you are tracking them. Those effective days are going to feel good, and that good feeling is going to motivate you to treat your time as the priceless commodity it is.

The Time Tracking Template is yours FREE forever. Once you complete this 31-Day Turnaround program, use the Time Tracker once a month for 3 months, then once per quarter forever to keep yourself accountable for how you spend time.

Today's challenge:

Get a clipboard, pen, and the Time Tracker Template set up and beside your bed for tomorrow. Track your time using a repeating timer on your phone every 15 minutes. At the end of your day, input the amount of sleep you are getting per night as the final entry. Two days from now, analyze your data and categorize your time: Sleeping, family/ friends, driving, chatting, screen time, phone screen time, actually doing work, doing important work, planning, etc. Draw your own conclusions, but look with a critical eye. Are you doing *important work* with your time?

31-Day Turnaround Time Tracker

Time	Activity	Urgent/Important, Non-Urgent/Important, Urgent/Not-Important, Non-Urgent/Not-Important
10:15	*Checking email*	*Urgent/Not-Important*
5:00 AM		UI, NU/I, U/NI, NU/NI
5:15 AM		UI, NU/I, U/NI, NU/NI
5:30 AM		UI, NU/I, U/NI, NU/NI
5:45 AM		UI, NU/I, U/NI, NU/NI
6:00 AM		UI, NU/I, U/NI, NU/NI
6:15 AM		UI, NU/I, U/NI, NU/NI
6:30 AM		UI, NU/I, U/NI, NU/NI
6:45 AM		UI, NU/I, U/NI, NU/NI
7:00 AM		UI, NU/I, U/NI, NU/NI
7:15 AM		UI, NU/I, U/NI, NU/NI
7:30 AM		UI, NU/I, U/NI, NU/NI
7:45 AM		UI, NU/I, U/NI, NU/NI
8:00 AM		UI, NU/I, U/NI, NU/NI
8:15 AM		UI, NU/I, U/NI, NU/NI
8:30 AM		UI, NU/I, U/NI, NU/NI
8:45 AM		UI, NU/I, U/NI, NU/NI
9:00 AM		UI, NU/I, U/NI, NU/NI
9:15 AM		UI, NU/I, U/NI, NU/NI
9:30 AM		UI, NU/I, U/NI, NU/NI
9:45 AM		UI, NU/I, U/NI, NU/NI
10:00 AM		UI, NU/I, U/NI, NU/NI
10:15 AM		UI, NU/I, U/NI, NU/NI
10:30 AM		UI, NU/I, U/NI, NU/NI
10:45 AM		UI, NU/I, U/NI, NU/NI
11:00 AM		UI, NU/I, U/NI, NU/NI
11:15 AM		UI, NU/I, U/NI, NU/NI
11:30 AM		UI, NU/I, U/NI, NU/NI
11:45 AM		UI, NU/I, U/NI, NU/NI
12:00 PM		UI, NU/I, U/NI, NU/NI

Copyright Hugh McPherson www.31DayTurnaround.com

Get this template and more FREE at
www.31DayTurnaround.com

Day 20:

Training Budget is Marketing Budget

I know that every hour and every dollar I spend on training my people counts as marketing my business. If my people are at their best, confident in their work, and I am confident in their abilities, our business will grow. I will dedicate specific time to training and retraining our public-facing staff, our support staff, and our production staff.

Marketing is sexy. We spend an incredible amount of time on marketing—obsessing over the right strategy, funneling thousands of dollars to ads, sending postcards, boosting Facebook posts, buying radio time—yet we always wonder, "Which marketing tactic is working?" It's no surprise we focus on marketing, because when we hit a big Fall Harvest Saturday and thousands of people arrive, cars flooding the parking lot, we feel great about ourselves.

Time is scarce for your customers. Think about how hard it is to convince a new guest to visit your business. There are a million other options on which people could spend their time. Family activities, youth sports, church obligations, jobs, and hectic schedules all compete for a family's time. Then, on the ONE DAY of the month that family can get out to do something, you need to be top of mind AND they need to choose YOU.

Training is marketing's multiplier: Good or Bad. Now, consider that family rolling into your parking lot, unloading and walking up to your admissions counter or walking over the threshold into your farm market. Then what happens? That customer is about to interact with your staff and at this critical moment, your level of staff training becomes a marketing multiplier.

Scenario Farm 1: The guests walk up to the counter and find your staff chatting amongst themselves, showing each other SnapChat videos and giggling. They are completely oblivious to the customers' arrival. In fact, the Mom has to say, *"Excuse me..."* Your staff looks up and sighs, putting their phones in their back pockets, and giving a half-hearted, *"How are you?"*

Scenario Farm 2: A family parks and the parking attendant tells them where to go for ticketing. At the front desk, the family receives a *"Hey there! Ready to Get Lost?"* They are explained the ticketing options and rung up quickly, then directed to *"put on those wristbands and head over to the orientation video. They'll tell you the games and the rules and you're off!"*

If Farm 1 & Farm 2 both spend $5,000 on advertising, which Farm will receive the highest Return-on-Investment (ROI)? Training is the marketing multiplier. If you spend time and money on training your staff, you can spend less on marketing; it multiplies the investment. Every guest you convince to visit, every client who gives your business a chance, comes away with a great experience to share.

If you neglect training, plan to spend a lot more on marketing. Negative experiences multiply your marketing, too. You will have to spend a lot more to convince new customers who *don't yet know* about the bad service they are about to receive. Word of Mouth marketing can actually work *against* your marketing budget.

Today's challenge:

Plot the key client, customer, or guest interaction points in your business. Write down exactly how your staff should interact, down to the scripting level, by thinking through how you or your best people handle those situations. If you are the best at these interactions, take careful mental notes as you handle a typical interaction, then write those notes down. Set aside a time within the next week to have specific training time with your staff away from guests or customers.

When a new school tour leader arrives for the first time, the team should...

Day 21:

"Little Thank Yous"

I commit to finding ways to spread 'Little Thank Yous' to individuals on the team any time they go above and beyond. I know I can't save my appreciation up for a year to distribute at an annual party. I will keep the 'Little Thank Yous' infrequent enough to keep them special, but frequent enough to keep people feeling good. I know that money isn't the key factor, but thoughtful, timely gifts directly connected to superior performance.

You can't save it all up. In our business, the holiday party is far removed from the busy season. So many times, the heroics of our people happen when we are busiest, too. By the time the Christmas party rolls around, it could be months since the heroic event and the excitement, and the positive feelings will have dissipated. Since you can't save it up, plan to spread it throughout the year with 'Little Thank Yous'.

Stock your *'Thank You'* basket. From any online company (we use VistaPrint), you can get Thank You notes preprinted with envelopes so you are ready to deploy a 'Thank You' at a moment's notice. Keep Starbucks gift cards (we stock $10 cards), local gas station cards ($10 is enough for a nice snack or lunch), and movie cards or steak house cards for bigger successes.

Strike whilst the iron is hot. If you start looking for opportunities, they will present themselves. Your mechanic was off on Saturday afternoon, but he came in special to tow your tractor from the field when you broke down. Your market manager sells the most pies ever in a week, AND didn't run out of varieties. Your bookkeeper finds a double payment and gets you $687.24 back from a vendor.

It's the connection, not the amount. The immediate connection from workplace 'awesome-sauce' performance, way above and beyond, is the key. It's your handwritten note that counts... the gift is merely a token.

Today's challenge:

Get a stack of thank you cards and a few local gift cards. Keep them ready in your desk drawer, so you can deploy them as needed. Look this week for ONE incredible performance that you can secretly reward with a handwritten note and a small gift—a 'Little Thank You'.

Day 22:

Don't Get Lazy

I will keep my eyes on the goal of improving my relationship with my staff. I will endeavor to keep my speech encouraging and uplifting. I will note and consciously 'feel' my feelings and reactions, before speaking to my team. I will vigilantly improve my own behavior towards my people. I understand that as I make a change in myself, others will sense that change, wait to see if the change is temporary, then, after time, see that I am working hard to improve.

Your brain is lazy. Your brain purposefully chooses the path of least resistance. This is self-preservation—your brain uses fewer calories when it can run on autopilot. Think about it. When you learned to drive, it was two hands on the wheel, triple-check all the stop signs, and constant focus to stay between the lines. Now, you likely *don't even think* about driving to work. You've done it so many times, you can talk on the phone, look at the scenery, adjust the radio, and correct your children... all while driving.

Habits are comfortable. We all have our comfortable habits and routines. I can make a hot breakfast for my kids and wife in the morning—3 breakfast burritos in 12 minutes—without a second thought. It's comfortable; it's a routine. When you go to the grocery store, I bet you shop the *same* store, the *same* aisles, in the *same* order, picking nearly the *same* foods every time. It's comfortable.

Your bad habits are comfortable, too. When you get frustrated, you yell obscenities. When the tractor won't work, you throw wrenches. When you can't remember a customer's email address, you get distracted and surf Facebook. Why? Because it's comfortable.

You are in the middle of a BIG change, and your brain hates changes. Don't get lazy. The minute you let your focus slip in this 31-Day Turnaround you have committed yourself to complete, you will drop the ball. Your old habits will rise up and threaten to overtake you, threaten to undo the progress you've made.

Only when this new way of working through challenges and working on relationships becomes so familiar, so routine, will you be able to relax and let the new, positive habit drive you to repeated, intentional behaviors.

Today's challenge:

Re-write the reasons why you began this 31-Day Turnaround program. Write down the negative experiences that drove you to improve. Re-focus on the new, successful outcome you are working to bring to your life and your business.

Time cards turned in late drove me NUTS and kept me up until midnight on Thursdays!

Day 23:

Speak Aloud Words of Encouragement

I will speak words of encouragement. I will start with my internal dialogue, changing my self-critical talk into internal encouragement. When I speak to myself, I will speak encouragement out loud. I will use encouraging words with my staff. I will be gracious and thankful, using kind words. I will slowly limit profanity in my own speech, knowing that encouraging words soften hearts and build up my people.

"I will not use foul language!" is recited by each group of guests as they view the video introduction on the way into the corn maze. Words are powerful. Written words have launched revolutions and religions. Words have brought tyrants to power and words have pulled them back down.

What you say each day matters. We put the rule *"I will not use foul language!"* in the introduction video *and* lead the guests in a rousing chorus repeating the phrase, because foul language is the gateway to bad behavior.

Think back to the last time someone dropped the 'F-Bomb' on you. What was your visceral reaction? You likely wanted fight or flight, but either way, your adrenaline started pumping. Words matter. Words create feelings and reactions.

I didn't make this idea up. It's been around since biblical times. *"Don't use foul or abusive language. Let everything you say be good and helpful, so that your words will be an encouragement to those who hear them."* Ephesians 4:29, New Living Translation

Think back to the last time someone dropped a genuine compliment on you. What was your visceral reaction? You almost immediately smiled and your cheeks flushed, then you likely tried to deflect the compliment with, "Well, it was nothing really." You felt warmth towards the compliment giver, maybe even relief because you know you had been trying hard to succeed or be kind or be helpful.

Author John C. Maxwell calls this the *"Elevator Principle"—We can lift people up or take people down with our relationships.* It all starts with your language, then expands to the language of your team.

If you slowly limit profanity in your speech and discourage profanity on your team, the whole team will feel the benefits. Profanity is never a positive addition to conversation. To build a positive environment with a lot less adrenaline pumping, limit profanity.

If you slowly work to incorporate words of encouragement into daily conversation with your people, they will slowly pick up the cue and share the encouragement. *"Nice work. Great job. Thank you. Couldn't do it without you. You're the man. You go, girl. You are killing it."* These simple phrases add positive energy to you and your crew.

It was hitting the afternoon rush at the farm market one October Saturday and I was exhausted. I pulled up outside and the line was stretching into the parking lot. I knew the crowd would be demanding our world-famous apple cider donuts. I snuck in the back of the bakery to see my crew feverishly mixing, measuring, frying, and packing donuts. I had run out of ideas for motivation, so I burst out with, *"Encouraging Words! I don't know what else to say, ladies, but I thought you'd need some, so... Encouraging Words!"*

Even my saying the words *"Encouraging Words!"* at the right time and with *feeling* had the effect of lifting spirits and communicating that I cared. Your people are *starving for encouragement* and they are starving for someone to care about them.

Don't you want to be married to someone who appreciates you? Wouldn't you want to work for someone who appreciates all you do? Don't you enjoy and savor each and every "Thank You" offered to you?

Become a healthy source of appreciation for your people.

Today's challenge:

Find some way to work *"Nice work. Great job. Thank you. Couldn't do it without you. You're the man. You go, girl. You are killing it,"* into conversations with your employees today. If you haven't been good at speaking your appreciation out loud, it will take a few tries to get it to naturally come out of your mouth. It might also take a few tries for your crew to believe it, but once they know you are genuinely appreciative, the effects will be positive and lasting.

Day 24:

You are Not a Born Leader

I understand that no person, man or woman, was born a leader. Leaders are self-made, and I will make myself into a more effective leader through incremental changes day by day. I am released from concern that I was not born to lead. I now actively choose to lead, and lead more effectively each day.

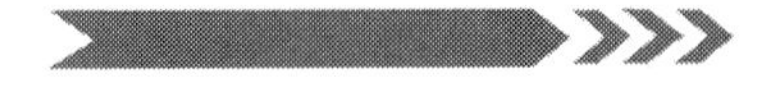

"Leaders are not born, they are always self-made." Steve Wiley of The Lincoln Leadership Institute at Gettysburg in his book, *Leadership Sessions from Gettysburg*, offers that effective leaders grant followers a stake in deciding the strategy for achieving the organization's goals. Purposefully involving others is a *learned skill.* Fortunately it is a skill that is available to everyone regardless of birth order, gender, I.Q. test scores, or sales commissions.

What a relief. With social media, it is easier than ever to make comparisons between yourself and other leaders. You may have a Facebook friend who seems to have the golden touch. Everything goes her way, at least on social media. You may know personally successful business leaders and think, *"I was just born deficient in the leadership department!"* What a relief to realize that leadership, true successful leadership is available to everyone through *behaviors*, not through birth.

Leadership is quality behavior over time. You have already learned, and hopefully put into practice, leadership skills over the past 24 days. You have already modified some of your behaviors, your speech patterns, and your actions to improve relationships with your people. You are working *right now* to improve yourself.

The fact that you are working on behaviors proves that you cannot be born with the skills to lead. You must *learn them* and *earn them*. The shackle of 'birthright leadership' is off! Now, also gone is *the excuse*.

You are not a 'born leader', and neither is anyone else. You *choose* to lead, and choose to lead well.

Today's challenge:

Write down your top three projects underway. List a few ways or areas of the project in which you could directly involve your team in deciding the strategy. Make sure your team is utilized as a resource for ideas, strategies, and timelines. You have a group of people whose continued success depends upon the business' continued success. They are invested in the outcome, so make time to invest them in the strategy.

Expanding the farm market, replanting the peaches...

Day 25:

Check Your Short-Term Memory

I choose to forget all past wrongs and mistakes perceived by me from my people. Especially, I'm choosing to let go of the long-time slights I've been holding on to for years. I choose to start each day with each person with a blank slate.

I remember it as a pivotal moment in my relationship with my dad. We had just purchased—after years of hassles with old, small stake-body trucks—a BIG, NEW farm truck. This gleaming white, 22-foot dump bed beauty was a revolution at our farm. It was a diesel truck, and by comparison all our previous, old gas-powered models looked like children's toys.

Even better, we had purchased a new equipment trailer that added another 24 feet of backhoe-carrying awesomeness to the rig. Air brakes, hydraulic dump, Roadmaster split shift, and dual axles on the truck AND on the trailer. It was a farmer's dream machine parked in our very own pole barn.

"Why don't you pull it out for Brad," my dad said.

I was in heaven. Being all of 16 years old, having cut my teeth with the old trucks hauling lime, I was ready for the upgrade. I climbed into the cab and fired it up. Sure, it had new instruments, but I'd seen the guys drive it and I had paid attention. I popped the air brakes and slowly pulled out of the barn with that beast purring like a kitten.

As I checked my mirrors and turned down the driveway, heading back to where my dad was standing, he and the guys were waving! Eventually, I realized it wasn't a parade-style wave, so, confused, I stopped the truck and hopped out.

The guys walked past the cab and back towards the trailer. **Someone whispered**, "You left the trailer brakes on..." One of our guys pulled the truck forward one half a tire rotation to show what I had done. You see, the truck was so powerful, I didn't know the trailer brakes were still engaged, locking the new trailer's brand new tires in place. My short trip from the building had left a flat spot on eight brand new tires that would have to be replaced.

What would you do? The guys were standing around—my dad's employees—and me, his son, who's supposed to be smart and efficient has just cost the farm a couple of thousand dollars on the new trailer, which now can't be used today.

I stood waiting for what I imagined would be quite a reprimand... that never came. *Ever.*

"Well, next time, you'll have to check the trailer brakes before you take off," said my dad. *"Pull it over to the shop."*

That was it. And that was it *forever*. I've told the story a dozen times as I'm sharing it with you, but my dad has never mentioned it. No yelling. No public humiliation. No memory of the incident.

How's your short-term memory? Are you still harboring grudges and mistrust of your people at work? Are they still secretly unforgiven in your mind and are you secretly 'lying in wait' for them to 'do it again'?

Today, any time I climb up into the cab of one of the big trucks with a trailer in tow, I always pop the air brakes, pop the trailer brakes and check for tires rolling before I head down the road.

Dad taught me that day how powerful it is to 'not remember'... to *not* yell, to *not* humiliate, even when he had every right to do that and worse.

He taught me the value of a short-term memory.

Today's challenge:

Write down your secretly harbored mistrust, your unforgiven slight next to the name of each person for whom you hold it.

Speak out loud, "I forgive you for _______________. Today we start with a clean slate."

Day 26:

Life in Congruence

I will bring my work life, my family life, and my personal life into congruence. I will be kind, gracious, and thoughtful with my workers, my family, and in my own internal voice. I will work diligently to care for my team as I care for my family, and as I take care of myself. I will be thoughtful of my team's physical and mental health, and their well-being. I will be considerate of my family's physical and mental health, and their well-being. I will purposefully be considerate of my own health and mental well-being. I will live life in congruence.

You know what imbalance feels like. Often during our busy seasons, we are operating at 100% at work, but we are neglecting our family life, purposefully sacrificing family temporarily for 'the good of the cause'. Whatever we name it, it is unhealthy and unbalanced.

You know what harmony feels like. We all have experienced moments of harmony. It is that wonderful day where everything goes well at work, your kids are well behaved, your favorite team wins the big game, and you get along great with your spouse.

Congruence is defined as 'agreement or harmony; compatibility'. When you are out of balance in one area of your life, it pulls the other areas out of balance, too. If you come home from work frustrated and angry, those feelings will bleed into your home life. If you are upset with your teenager, it will be hard to concentrate at work. If you are struggling under a load of business debt, wondering each day if you'll be able to make payments and financially survive, you'll be mentally absent when you play with your kids.

If you have let your physical health go, your self-image suffers and you lack energy. If you are not speaking to your mother because of *'what she said at Christmas last year'*, your mind is preoccupied with that disagreement. If you are drinking too much after work, your health and all your relationships are in jeopardy; they are incompatible.

Bring your work, life, health, and personal relationships into harmony. In Day #8, we talked about 'Falling on your sword' as a metaphor for being the one who takes responsibility for fixing damaged relationships.

Until you bring harmony throughout all the facets of your life, you are struggling under burdens you do not need to carry. Congruence—harmony—will lighten your load.

Today's challenge:

Under each topic below—work, life, health, and personal relationships—list stresses you are feeling right now. Write a few action steps for each that could directly work to eliminate the disharmony. If you were to take full responsibility, pick ONE stressor you could eliminate and bring into congruence today, then take immediate action, even if it is just a first step, to make it happen.

Work

___

___

___

___

Life

___

___

Health

Personal Relationships

Day 27:

Stop Living at the Redline – Part 1

I understand that maxed-out life at the redline is unhealthy. I consciously choose to evaluate the projects in my work life and select only the highest-value projects to continue. I want to live a balanced, exciting life without the stress of pushing to the limit each day. I will deliver maximum effort during short periods of time to close open loops and complete projects by deadlines, but I will not live every day of my life at that extreme. My mind, body, and sanity are valuable commodities I will treat with respect. I choose to stop living life at the redline.

The Redline is that point on an engine's tachometer that indicates the speed of the engine is increasing to dangerous levels, generating excess heat, wear and safety concerns. Continued operation with high RPMs (revolutions per minute) at or above the redline will damage the engine.

Extreme effort. 2015 was a whirlwind year for our business and for me personally. One in which, if I had ever wondered the limits of what I could mentally and physically accomplish in one year, I found out. In addition to running our home agritourism attraction—pick-your-own, school tours and farm market, designing games, corn mazes, and teaching online classes and shipping products to farms all around the world—we took a leap to launch a winery, produce our first wines, build a tasting room, and complete all our licensing. We started February 1st with the paperwork and served guests December 4th. It took *extreme* effort.

Maybe you, too, have found your limits this year or in a past season of life. No matter how productive you were in those times of extreme effort, you likely found yourself exhausted, a feeling of things slipping between your fingers like sand.

That is what I call *Life at the Redline* and **this is not a heroic tale, but a cautionary one.** While I survived that year and much was accomplished, I wondered at what price? Are you living 'Life at the Redline'?

There is a downside to optimism. Though shocking for you to hear me write these words, hopefully you know optimism is my natural state of being. The downside is that the world is full of possibilities and, as an optimist, you feel like most of them are possible for you, because, "Why wouldn't things work out?" If everything is likely to work out, why not do everything?

Optimism and impatience can be deadly. This dynamic duo of traits leads you to false conclusions such as, "If everything is going to work out, why not do it all—now?" Impatience demands that you do everything, and optimism says it won't take very long to do. If it won't take very long to do, you should add it to be done today. If you just stay a few more hours, you could get a lot more done, and so on, and so on.

Have you ever had the experience of scheduling your day to the minute, then rushing from activity to activity without enjoying any of them?

Marry a practical girl. My dear, sweet wife Janine is a practical girl. She knows exactly when I'm leaping off the deep end, and she'll warn me. However, optimists brush aside such warnings as 'negativity', and seldom heed them. I learned that year that she was right. I was 'redlining' myself as project after project got added to the list. She was, indccd, right. She had a keen insight from outside my 'sparkling world' of projects. I was pushing towards the Redline.

Are optimists doomed? No way. Optimists rule. If you lose your positive attitude, things go downhill quickly. I submit that you should purposefully work to be *more* optimistic: Smile more. Have faith it will work out. Believe the best in people. Delegate and trust. Nothing can sustain you through hard times like a positive attitude. Your attitude will lift others up as well.

Who is your practical guy or girl? Who watches from the outside to see if you are overloading your plate? To whom are you accountable?

Today's challenge:

Evaluate yourself. On a scale of 1 to 10, 'ONE' being sitting on a beach relaxing, and 'TEN' being 12-hour-days/7-days-per-week, where are you *right now* in your work life? If you are a high-achiever, it is quite possible you generally live between 7 and 9 on the scale.

Next, print out a blank calendar. Looking back on the past 3 months, by week, rate your mental state on the same 1-10 scale. How many weeks out of the past 3 months have you been at the Redline (9-10)?

Day 28:

Evaluate Your Redline – Part 2

I understand that maxed-out life at the redline is unhealthy. I consciously choose to evaluate my Achievements, my Finances, and my Emotional State separately. I will delve into each area of my life to honestly assess and plan actions to improve each, step by step, over time. I choose to stop living life at the Redline.

Different Redlines. Seldom do you find yourself at the Redline in all aspects of your life. There are different Redlines, and like me, you've likely seen the edge of each at some time in your life.

The Achievement Redline. It's your schoolwork or your work-work or your work on your dreams, but it's what you are working on that's pushing you to the limit. Your job, your boss, your people, your desire to succeed makes you take on more projects, more classes, more responsibilities. You feel like you can do it all and you can get it done, IF you just work a *little* longer, a *little* harder, on a *little* less sleep.

How the Achievement Redline hurts you. Typically, pushing the Achievement Redline leads to multi-tasking. You try to do it all at once which leads to inefficiency. I found myself constantly interrupted, having to switch back and forth from project to project, losing time with every switch. When you are short on time, you miss things—inefficient parts runs, stuck with the vendor who can do it *now*, not the *best* vendor for the job.

The Financial Redline. Most business owners go through the Financial Redline when they 'bank on the future', rob Peter to make payroll, delay Paul to buy fertilizer, and don't pay yourself because 'it'll all come back as profit in the end'.

How the Financial Redline hurts you. I've never been at the financial redline without missing an opportunity because resources were locked up. It's nerve-racking when you don't have a financial cushion for regular operating expenses, and that takes a toll on your critical thinking; it puts you on edge. Jobs typically get underfunded, so they cost nearly the same as 'done right', but are low-quality and likely will need to be replaced.

The Emotional Redline. Back in 2005, my mother was diagnosed with lung cancer. She fought for a year and a half, but she didn't make it. She was the heart of Maple Lawn Farms and so much more to us, to our church, to her civic and agricultural groups. During that time, I was running our business, growing my very young family, and *living* at the Emotional Redline for two years.

How the Emotional Redline hurts you. *Well, your hair can fall out.* I found that out the hard way! I can share with you that I was not the best boss, husband, father, son, business partner, or anything during those years. Business is stressful enough. It was faith and time, a patient wife, and my kids that slowly brought me back from the Emotional Redline.

Disciplined thankfulness brings you back from the Redline. If you are at the Financial Redline or the Emotional Redline or the Achievement Redline, it can be hard to be thankful, but *hard* is different from *impossible*. Disciplined thankfulness is the hard work done to be thankful for in the midst of life's storm.

Disciplined thankfulness is a personal journey. Here are some examples:

Let someone off the hook. *Forgive her.*

Use positive, encouraging words with your people. *Be peaceful.*

Pause, and look into the eyes of a loved one. *Savor the joy.*

Hold on a little longer when you hug Aunt Bertha and say, *"I love you,"* to your parent, to your child.

Make one extra payment. Be fruitful.

If life is hard right now, be *disciplined* in your thankfulness. Slowly, you'll return from the Redline.

Today's challenge:

Evaluate your Achievement, Financial, and Emotional Redlines. Write down the stresses you feel each Redline places upon your life. Brainstorm 20 things for which you are thankful and write them down. Now brainstorm 20 'Disciplined Thankfulness' actions you could take to evoke positive change that moves you away from *your* Redline.

Achievement

___

___

___

Financial

___

___

____

Emotional

____

____

____

I'm Thankful For:

1______________________________________

2 ______________________________________

3 ______________________________________

4 ______________________________________

5 ______________________________________

6 ______________________________________

7 ______________________________________

8 ______________________________________

9 ______________________________________

10______________________________________

11 ______________________________________

12______________________________________

13 ______________________________

14 ______________________________

15______________________________

16 ______________________________

17______________________________

18 ______________________________

19 ______________________________

20 ______________________________

Thankfulness Actions I Can Take:

1 ______________________________

2 ______________________________

3 ______________________________

4 ______________________________

5 ______________________________

6 ______________________________

7 ______________________________

8 ______________________________

9 ______________________________

10 ____

11 ____

12 ____

13 ____

14 ____

15 ____

16 ____

17 ____

18 ____

19 ____

20 ____

Day 29:

The Redline Is Temporary – Part 3

I understand that maxed-out life
at the redline is unhealthy. I
understand that pain is
temporary and that, if I have to
go through pain, I might as well
get something from it:
I will push through
and WIN.

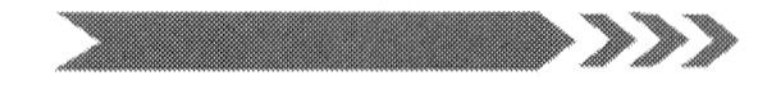

Pain is temporary. I found this in an Eric Thomas (known as "ET, The Hip-Hop Preacher") motivational speech years and years ago when I was struggling. I quote it. I make my kids listen to it. The next time I'm asked to speak at a high school, I'm leading with it.

Pain is temporary.

Temporary Redlines. Redlines should be temporary, too. You might be at the Achievement Redline as you crush it in October, but then you back off. You might take your farm to the Financial Redline for a major expansion project, but then you work on sales growth and payback. You might experience trauma at the Emotional Redline, but you don't build your house there. You feel it, you grieve, you move on. Redlines are meant to be temporary.

Get something from it. Mr. Thomas encourages that if you are in pain you should get something from it! If you are working 60-hour weeks, make sure you are growing your business. If you are taking evening Master Classes, do the work and get an "A"! If you join our Boot Camp, buckle down, do your homework and get a system in place for next year. If you are figuring out how to deal with a teenager, list the things you love about your child. Whatever it is, the pain is temporary. You might as well get something from it.

Pushed to the max. In 2015, we were pushing the Achievement Redline to the max. We tackled so many projects I was maxed out managing the decisions and details. *It was painful*, but from it we emerged with the stage set for a much different 2016. We did NOT take on major building products, major product launches, major investments. We decided 2016 was the year to market, cross-market, and grow sales to pay for everything we had done the previous year. We pushed hard, but we got something from it.

Push to the max, then back off. 2016 saw us back away from the Achievement Redline. We had plenty of challenges incorporating the new products and attractions into a cohesive offering for our guests. The team needed a breather from the breakneck pace of the previous year.

Plan to Win. Joe Calhoun in his book *On the Same Page* says, *"Every hour spent planning is worth four hours working."* It's so tempting to just keep working, figuring that everything else is 'going to work itself out'. It won't. You'll end up tired, worn out, and no further ahead.

Block out time to think, to plan. It's nice to have a big dream, but, as Antoine de Saint-Exupéry said:

"A goal without a plan is just a wish."

The courage to think. It takes courage to stop doing things. Step back and think about where you want to be, who you want to be. It takes courage to shut down something that you've done for years, but isn't making money. It takes courage to pick a new direction to move definitively toward.

Continue to grow. Invest time and money in growing yourself and growing your team through a detailed planning process. Every hour you spend planning your projects and planning your success is worth four hours of actual work.

Take the time—*make the time*—to pre-plan your success, and maybe you won't Redline along the way.

Today's challenge:

Pull out your calendar and block off ONE FULL DAY within the next 14 days to spend alone planning. Dedicate yourself to planning first, then plan a second day, within a week of your solo planning day, to include the team. Your team needs to see the value *you* place on planning for the business, so they know their time spent planning *with you* is valued *by you.*

Day 30:

Public Recognition for Your People

I know it is not all about me. I will purposefully, publicly recognize the contributions of my staff informally today, and I will plan to recognize them formally in the near future. I cannot do it alone, and as a leader I choose to elevate those around me, knowing that their successes are ultimately successes for the business.

Over the past 30 days, you have made incremental changes to the way you interact with your people and in how you manage yourself. **You deserve a lot of credit for embarking on, and sticking to this 31-Day Turnaround program.** That credit you get from me and you can give to yourself.

Your people need credit given to them from you. You need to publicly reward the work they do for you and for the company. You need to consciously give them the credit for the job they do.

How do you know you need to do this? Because you have a desire inside you to receive public credit, public acknowledgment of your efforts, you want exactly the same thing. Modesty aside, you want credit from those around you.

In one of life's great paradoxes, the only way to receive public credit is to give it away. The final linchpin in your 31-Day Turnaround is to freely give your people the credit they have earned and that they crave.

We hold an annual Holiday dinner where we pick up the tab and at which we celebrate our employees. **For our teenage and youth employees,** it's an unlimited pizza buffet followed by a short program at which I call each employee up front, publicly thank them for their individual contribution, and give them a silly certificate that directly correlates with their contributions from the season.

For our adult workers, we take them to a local, sit-down restaurant, rent out a room and let them order anything they want off the menu, including drinks. The adults, too, are called individually in front of the group and we present them with a silly certificate specific to each person while thanking them for their contribution to the business.

Simple? Yes. Effective? You bet. We all crave recognition for the effort we put forth. Instead of withholding it, lavish it on your people. *They could work anywhere*, yet they *choose* to work for you, with you.

It's time to publicly say, *"Thank You!"*

Today's challenge:

In your regular day today, pick at least two people on your team to say, *"Thank you, I don't know what we'd do without you."* Do it TODAY. Begin planning an official season-ending party or holiday party at which you will publicly thank each employee separately with fun, positive, yet silly certificates.

Day 31:

Maintain Your Irrational Belief

I have such faith in my ability to create positive change in my life that I am unshakeable. I know I will be challenged along the way, as I enact changes in my behavior. I know I will be challenged by people on my team who won't believe these changes are real or lasting. I know I will be tempted to return to my old habits. I choose to stay focused. I choose to stay on course. I choose to disprove the doubters. I gather strength and energy to continue my irrational belief in this new world of work I am creating step by step and day by day.

Being 'rational' is a euphemism for being 'average'. The world is full of average. There are so many people in this world ready to bring you down and talk you out of improving and rain on your parade, you'd have to be irrational to survive. In a world full of average performers, your belief in yourself must be irrationally strong.

A world of critics. It sure is easy to be a critic today. Yelp, Facebook, Google Reviews, Instagram and TripAdvisor let anyone in the world write a scathing review of you or your business. Try 'running an idea by someone'. Inevitably the response will be, "Well that's a neat idea, but... [INSERT NEGATIVE CRITIQUE HERE]."

It will be presented by the person as *'thoughtful feedback'* or *'just looking out for you'* or *'well, not many books get published'*, but it is simply 'Idea Murder' brought to you by someone without a good idea of his own.

You are not average. You are already successful, and getting better. Your irrational belief that this 31-Day Turnaround would work, would create positive change in your working relationship, has kept you working through the program for 30 days!

Simply sticking to the program for this long *validates your irrational belief* in your own fortitude and motivation.

Today's challenge:

On the following page, write down any experiences from the past 30 days that have already been successful for you or have indicated positive progress. In the second list, add the name of anyone who heaped / is heaping / will heap negativity on you when you describe the 31-Day Turnaround you are working through. Continue to update each list moving forward.

Things that have Already Improved:

My team knows that I'm trying!

People I Will Prove or Have Proven Wrong:

Frank didn't believe my changes were real, until he got his ONE tool: A new electric grease gun!

What to Do Next

To put it simply – something hurts. You are reading this and looking for books that will help you have more fun with your team *and* get more DONE each day with fewer hassles. You know something needs to change, but you may not know what.

If you operate a farm, market, Agritourism destination, campground, swimming pool or other seasonal business with part-time employees, chances are good you need a better system.

That's why we created Agritourism Manager Boot Camp.

Boot Camp gives you the precise blueprint for a system that helps you and your employees get on the same page, manage expectations, resolve conflicts and get more DONE, together → Not just *ideas*, but *real* templates and materials you can use RIGHT NOW.

...PLUS, as a book buyer, you can get SAVINGS up to $300 on your Boot Camp Membership when you opt-in for the online extras.

Sign-up at www.31DayTurnaround.com/finished, for a special video message and more details about taking the next step with Boot Camp.

Thanks for reading. Don't forget to join our VIP Facebook Group and get support and guidance from other readers as you continue to improve the way you work in your business.

Start each day using the tools you learned over the past 31 days. Return to this book and reactivate your change-mindset any time and most of all, I wish you a new, productive and FUN season filled with success and profitability.

–Hugh

In their own words. Here's what farmers who have invested in Agritourism Manager Boot Camp have to say about the program and, most importantly, the streamlining it brought to their organizations.

"We have all our stations finished for our store, food shacks, etc. Now just working on the 'October only' activities! We have never been this farm ahead!"

– ANNIE BARTALEMIA, Lee Farms, OR

"We implemented the Maize Quest staffing system and staffing was smoother than ever. It really helped!"

– CINDY & BARRY BROEKHUIZEN
Harvest Moon Farm, MI

"We have found the course to be great so far! I think it will really help us get the right direction for our farm."

– SCOTT SKELLY, Skelly's Farm Market, WI

"I can remember Hugh's first visit like yesterday. Teaming up with Hugh and Maize Quest has impacted everything I do in a very positive way. Ag entertainment is taking my business to a level I never thought possible."

– MIKE MARINI, Marini Farms, MA

"Just hosted our first day of interviews. It worked so much better than our old set of questions."

– RACHELLE WEGELE, Anderson Farms

"It just seems that if we tweak your {content} to meet our needs we spend way less time than reinventing the wheel."

– LAURIE BUCKELEW, Buckelew Farm

"Agritourism Managers Boot Camp has been one of the best investments I've made for our business. We discovered we are not alone in our struggles as managers. ATMB gave us the tools we needed to make hiring and training much more streamlined. We've already implemented what we learned and simply couldn't be more pleased with our success!"

– AMY LADD, Lucky Ladd Farms

"I am also working on Open/Close Lists – great idea!! Hugh, thank you for sharing your knowledge! I am excited to put all that I learned to the test! I look forward to meeting you at your farm this summer!"

– RHONDA HEAL, Roba Family Farms

"Here is my homework from the last two modules. I'm being the copy and paste guy on this lol. What Hugh is using fits right in here! #copyandpaste"

– ANDREW DIXON, Granddaddy's Farm

"A lot of the basic work we did last year after my daughter came back from a conference and declared, "We have to get organized with our labor." I have to admit I didn't see how all this paperwork would help... it did a little last season, but now with this course, the way you present the material to the employees in such a straightforward, in-your-face, look at this everyday kind of attitude, I can see how helpful it is going to be. We are re-writing our job descriptions to the format you presented so they have no 'I didn't know I had to do that' excuses. Thanks for making the material and all the 'paperwork' make sense to me."

– MELINDA VIZCARRA, Vizcarra Vineyards

About the Author

Hugh McPherson, Maize Quest's Maze Master, has been "losing" guests and telling "corny" jokes since 1997 on his home farm. Maize Quest's Corn Maze & Fun Park is home to the annual corn maze, and now features the annual Sunflower Festival.

The home farm, Maple Lawn Farms, welcomes guests to pick-their-own peaches, apples, pumpkins & blueberries throughout the season. In 2015, Hugh launched Maple Lawn Winery to add value to the fruit crop and welcome a new tribe of wine-loving guests.

Maize Quest, through www.MazeCatalog.com, designs corn mazes, games, and attractions for over 85 farms in the U.S., Canada, and the U.K. The Agritourism Manager Boot Camp is Maize Quest's first online course. Having developed and trained over 800 employees, this system works with everyone from Grandma to a 14 year old in his first job. Boot Camp is the #1 program for agritourism and farms with seasonal, retail employees to build a complete, staff management system.

Hugh served for 6 years on the North American Farmer Direct Marketing Association's Board of Directors (NAFDMA). Hugh also serves as the choir director at Centre Presbyterian Church in New Park, PA, and plays Ultimate Frisbee in a local league. His pride and joy are his wife Janine, daughter Annie, and son Ian.

Made in the USA
Columbia, SC
19 January 2018